CONTENTS

WHAT'S A DESERT?

A desert is a very dry place. A place is called a desert if it receives less than 10 inches (25 centimeters) of rain a year. Many people believe that all deserts are very hot. However, there are also deserts that are very cold. Antarctica is considered a desert.

Some deserts are found near mountains. Wet air brings rain. The mountains block wet air. By the time the air reaches the area beyond the mountains, there is not enough water left to create rain. No rain causes a desert to form. This is how rain-shadow deserts are created. The Great Basin Desert in the western United States is an example of a rain-shadow desert. It lies east of the Sierra Nevada Mountains. Those mountains block the rain, making the land dry.

How Do You Live There?

LIVING IN THE DESERT

Joanne Mattern

PowerKiDS press.

NEW YORK

Published in 2021 by The Rosen Publishing Group, Inc.
29 East 21st Street, New York, NY 10010

Editor: Kristen Susienka
Designer: Rachel Rising

Photo Credits: Cover, p.1 Ikpro/Shutterstock.com; Back Cover, pp. 3,4,6,7,8,10,11,12,14,16,18,20,22,24,25,26, 28,30,31,32 (background) Yevhenii Borshosh/Shutterstock.com; p. 5 Macca Sherifi/Shutterstock.com; p. 7 Volodymyr Goinyk/Shutterstock.com; p.9 Homo Cosmicos/Shutterstock.com; p. 10 Johnny Adolphson/Shutterstock.com; p. 11 Seleznev Oleg/Shutterstock.com; p. 12 Tim Roberts Photography/Shutterstock.com; p. 13 Bettmann/Contributor/ Getty Images; p. 15 Christian Vinces/Shutterstock.com; p. 16 vichie81/Shutterstock.com; p. 17 Darren J. Bradley/ Shutterstock.com; p. 19 abriendomundo/Shutterstock.com; p. 21 John Theodor/Shutterstock.com; p. 23 Everett Historical/Shutterstock.com; p. 25 lavizzara/Shutterstock.com; p. 27 Sid0601/Shutterstock.com; p. 29 Songquan Deng/Shutterstock.com.

Cataloging-in-Publication Data

Names: Mattern, Joanne.
Title: Living in the desert / Joanne Mattern.
Description: New York : PowerKids Press, 2021. | Series: How do you live there?! | Includes glossary and index.
Identifiers: ISBN 9781725316409 (pbk.) | ISBN 9781725316423 (library bound) | ISBN 9781725316416 (6 pack)
Subjects: LCSH: Desert people--Juvenile literature. | Deserts--Juvenile literature. | Desert ecology--Juvenile literature.
Classification: LCC GN390.M374 2021 | DDC 306.0915'4--dc23

Manufactured in the United States of America

Some of the images in this book illustrate individuals who are models. The depictions do not imply actual situations or events.

CPSIA Compliance Information: Batch #CSPK20. For further information contact Rosen Publishing, New York, New York at 1-800-237-9932.

Some deserts are located on the coasts, near water. The Atacama Desert in Chile is a coastal desert. Strong winds blow water away from the land.

Wind can also create deserts. Most of the world's biggest deserts were created by trade winds. Trade winds are gusts of dry air that blow away clouds. This allows more sunlight to heat up the land. Africa's Sahara Desert is a trade-wind desert.

Some deserts are full of sand. Africa's Sahara Desert and Australia's Simpson Desert are sandy. Other deserts are rocky with few plants or animals, like Asia's Gobi Desert. The Gobi Desert covers most of Mongolia. It stretches for about 500,000 square miles (1.3 million square kilometers). The name "Gobi" means "waterless place" in Mongolian. Even though it has harsh conditions, people live in the Gobi Desert. They herd cattle and other animals. The people who live in the Gobi Desert are **nomads**. They move around to find water and food for their herds.

Antarctica is the largest desert in the world. You will not find any sand there, though. Antarctica is covered with ice all year long.

Desert Wildlife

Even though it's dry and empty, many animals live in the desert. In the Karakum Desert in central Asia, there are desert sparrows and corsac foxes. These creatures live in cacti or holes they dig in the ground. In Antarctica, there are penguins. Emperor penguins huddle together for warmth in strong winter winds. Some plants live in deserts, too, and so do people. Many living things have learned to live in these harsh places.

EARLY LIVING IN THE DESERT

Without water, no one can survive in the desert. Over time, people have come up with special ways to find water. Some of these ways are still used today.

In the past, desert people learned to survive with little rain. The Nabataeans lived in the Arabian Desert in what is now Jordan more than 2,000 years ago. Their main source of water was a **spring** called Ain Musa. The Nabataeans dug paths in the mountains for water to flow from Ain Musa to the city of Petra.

The Nabataeans also used **terrace farming** to **irrigate** their crops. Farmers planted crops along narrow valleys called wadis. Planting here allowed rainwater to flow down the slopes of the valleys to the crops.

Here is an example of ancient terrace farming, at a location in Yemen.

OASES

Some desert areas have underground springs. These springs help plants and crops to grow in areas called oases. These lush places were important stops on ancient trade routes. Many ancient and modern cities were built near oases.

Many deserts are very hot during the day. When the sun sets, it can get very cold. The ancient Ancestral Pueblo (or Anasazi) people figured out how to deal with these big temperature changes. They lived in the Sonoran Desert, which covers the southwestern United States and northern Mexico. They built homes in caves on the sides of rocky cliffs (see image below). The cliff walls kept them shaded from the hot sun, while the stone and mud homes stayed at a good temperature, whether it was hot or cold outside.

When the nomads move, they pack up their tents and bring them along.

Today, many nomads live in tents. The tents are made of thick cloth to keep out blowing wind and sand. Openings let breezes inside. The tents are usually light colors. These colors reflect, or shoot back, sunlight and keep the inside cool.

ANCIENT GROUPS

It can be hard to find your way in sandy deserts. The wind changes the shape of sand dunes and covers up footprints. Bedouins have lived in the African desert for thousands of years. They used the positions of the sun and the stars to find their way. Modern Bedouins still use these methods today.

The Sonoran Desert People lived in what is now Arizona more than 5,000 years ago. They used water that flowed down the mountains to water their crops. As the area became drier and the population grew, they had to change. About 1,500 years ago, they started digging **canals** to bring water to their villages. These canals were used for 1,000 years and can be found all over Arizona.

This modern canal is located in the state of Arizona.

The North American desert was home to many ancient people, including the Papagos, who lived near Mexico.

MODERN LIFE IN THE DESERT

Today, about 1 billion people live in deserts around the world. Like people in ancient times, these people need to find ways to locate water and live with extreme temperatures. Modern technologies have helped bring water to many desert communities.

Some deserts have water very deep under the ground. These water stores are called **aquifers**. In the past, it was hard for people to reach this water. Many groups didn't have the tools needed to bring the water up from the ground. Today, companies use powerful machines to drill deep underground. The water is then pumped to the surface. Pipes carry it to communities all over the desert.

Some cities in the California deserts get their water from rivers. Water also comes from snow melting off the Sierra Nevada Mountains.

Lima, Peru, is built on a desert where there is almost no rain. People get water from rivers and the ground, but there may not be enough for everyone. ⟶

MOVING WATER FOR PEOPLE

Another way to get water in the desert is to bring it there from somewhere else. Los Angeles in California is a large and dry city. It gets its water from canals, pipes, and concrete structures called aqueducts. The Colorado River Aqueduct is 240 miles (386 km) long. It stretches from Lake Havasu in Arizona to Los Angeles.

People who live in the desert also need to get food. Some food is brought in by truck or airplane. However, modern methods of irrigation also allow food to grow in the desert. Some irrigation systems use water delivered from other places. Others use water from aquifers. Water flows to the crops through a system of pipes and sprinklers.

The California Aqueduct, like the Colorado River Aqueduct, helps bring water to Los Angeles. Water travels through it for 444 miles (715 km) to reach the city.

STAYING COOL

Today, many desert buildings are designed to keep the heat outside. Many are painted in light colors. Electricity has also made it easier for people to cool their homes. Fans are used to create a breeze. Refrigerators and freezers keep food cold.

Now air conditioning also helps people stay cool inside. Public places like stores, movie theaters, and offices are kept at a comfortable temperature, no matter how hot it is outside.

On very hot days, many people use their air conditioners. However, air conditioners use a lot of power. Sometimes too many people run their air conditioning at the same time. More electricity is needed than power companies can provide. As a result, the power can go out for short or long periods of time.

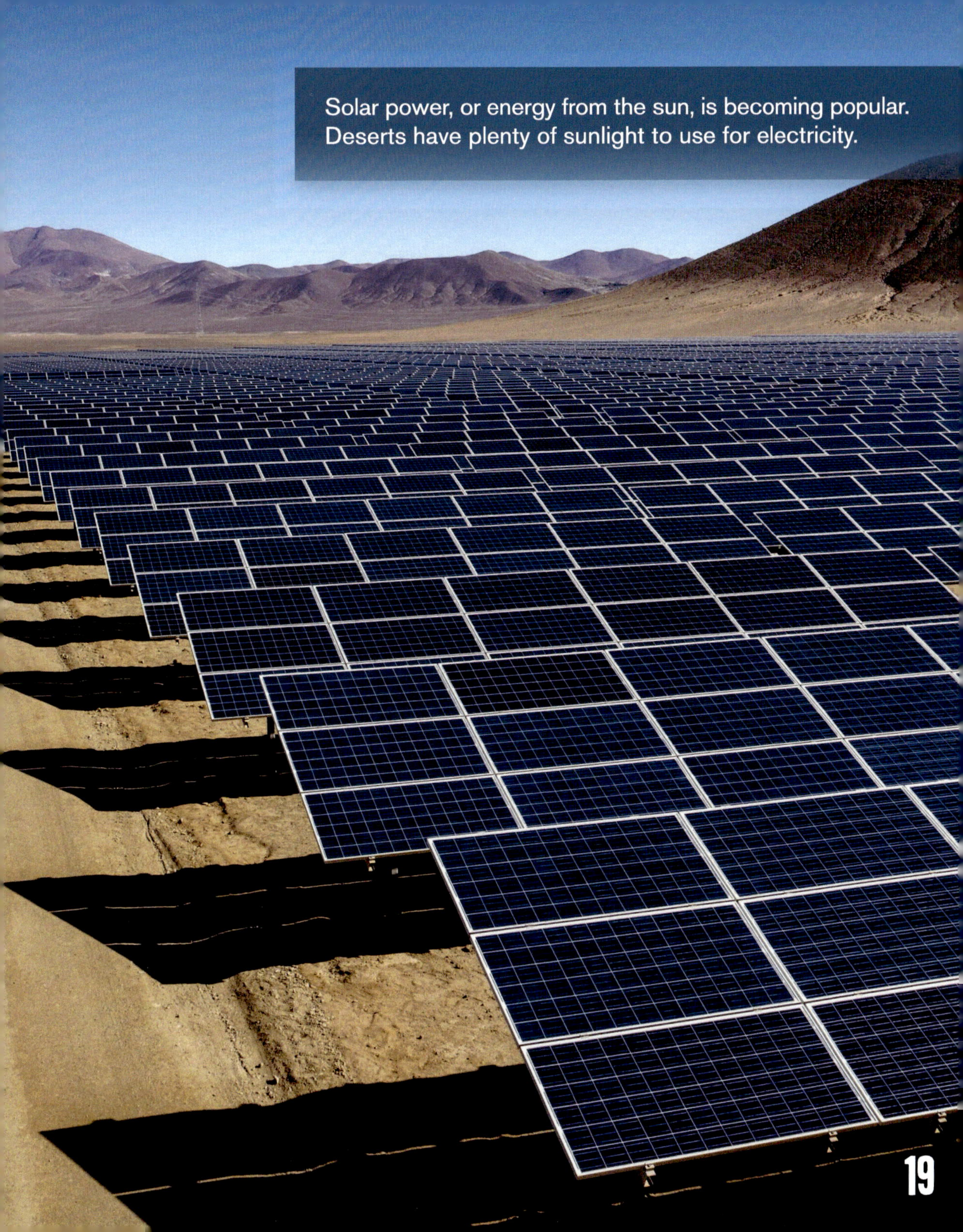

Solar power, or energy from the sun, is becoming popular. Deserts have plenty of sunlight to use for electricity.

People who live in the Sahara and Arabian Deserts often wear long, loose-fitting, white robes. These clothes allow air to move around the body. The white color reflects sunlight away from the person. Clothing also protects them from blowing sand and strong winds. Many people also cover their heads with a white cloth called a kaffiyeh.

In the deserts of the United States, cattle ranchers keep cool by wearing large-brimmed cowboy hats. Sometimes they wear light clothing. They use bandanas to help wipe sweat off their skin.

In Antarctica, researchers who study wildlife and climate wear large coats called parkas, special windproof gloves, and hats or earmuffs. They also have thick boots and wear many layers of clothing so they don't freeze when they go outside.

Kaffiyehs, such as the one worn by this Bedouin trader, protect the head from blowing sand.

CHANGING DESERTS

Deserts have always been on Earth. However, these places face new challenges because of the way people live. Today, deserts are spreading. This is called desertification. Destroying **natural resources** is a big cause of desertification. Grass and other plants store water and hold the soil in place. Removing grass and plants makes the soil dry up and blow away.

People also cause problems when they cut down forests. This also makes soil dry out and blow away. During the 1930s, **overgrazing** and poor use of the soil hurt the central part of the United States. The area was named the Dust Bowl. Winds made the soil blow away, creating huge dust storms. Millions of people lost their homes.

Strong winds blew dust across the American Midwest during the 1930s, burying houses, farms, trees, and more. ⟶

WATER SHORTAGES

As cities grow bigger and more people live in desert areas, they need more water. In some places, water supplies run out. Several countries in Africa get water from Lake Chad. It is located on the edge of the Sahara Desert. Because it is near the desert, the lake has shrunk to less than half its original size in the past 50 years.

Climate change is also affecting water amounts throughout the world. As temperatures rise, weather patterns change, leading to both fiercer storms and longer dry spells. Some areas that saw lots of rain in the past might not see as much today. Other areas that received a lot of rain might suffer droughts now. Places experiencing droughts can learn about surviving with less water from desert communities.

This view of Lake Chad from space shows how the water is drying up and making the lake smaller.

Mojave Desert Water

The Colorado River in the United States is the source of water for seven states: Arizona, California, Colorado, Nevada, New Mexico, Utah, and Wyoming. Today, it faces challenges that make providing water to all the cities that need it more difficult. One challenge is that the river dries out. A solution to this problem is to look elsewhere for water. The nearby Mojave Desert has large stores of water below it, but people are unsure about using the water there. They worry using the desert's water might threaten, or cause harm to, wildlife living in the desert.

FIGHTING BACK

Governments around the world are fighting back against the spread of deserts. One way to help is by planting trees. Communities plant trees in rows called windbreaks. Windbreaks stop the wind and help hold soil in place.

Trees aren't the only way to create windbreaks. Workers in China's Tengger Desert have built straw fences along the sand dunes. These fences stop strong winds and keep sand from blowing away. This means the desert cannot grow bigger.

Governments also educate people on better ways of farming and raising cattle. They want people to become more aware of natural resources and the best ways to use them. We can keep deserts from spreading. We can also help people who live in deserts live better, safer lives.

Windbreaks, such as this row of trees, keep soil from blowing away, which helps prevent desertification.

HOW SCIENCE AND YOU CAN HELP

Scientists are developing exciting new ways to keep deserts from spreading and to bring water to dry places. One idea is nanoclay that can be sprayed on desert sands. Nanoclay keeps the sand wet. This makes it clump, or bunch, together and not blow away.

Other scientists have found a way to take water from air. Even dry desert air has some moisture in it. In 2018, scientists at a university, or a school after high school, called MIT made a **device** that takes the moisture from dry air. If this device became more common and larger, it could help cities and homes in the desert have water all the time.

You can help the desert **dilemma** too. You could think of a new device to help create water. Or you could write to people in government and ask them to make good decisions about climate change.

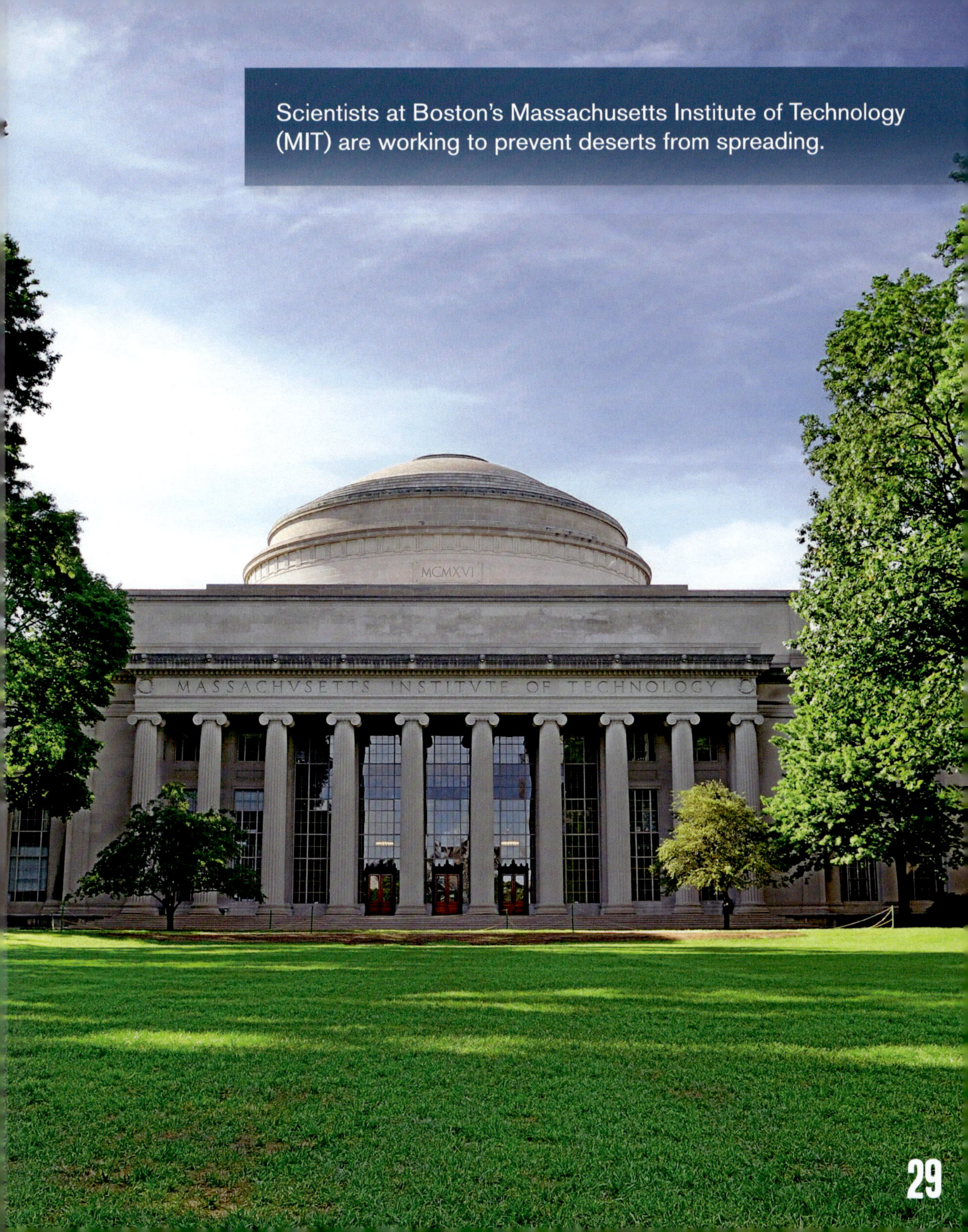

Scientists at Boston's Massachusetts Institute of Technology (MIT) are working to prevent deserts from spreading.

MAJOR WORLD DESERTS

DESERT NAME	AREA	CONTINENT	RAINFALL PER YEAR
Antarctica	5.4 million sq mi	Antarctica	< 2 inches
Sahara	3.3 million sq mi	Africa	< 0.9 inch
Arabian	900,000 sq mi	Asia	< 4 inches
Gobi	500,000 sq mi	Asia	2–8 inches
Patagonian	260,000 sq mi	South America	6–8 inches
Great Victoria	250,000 sq mi	Australia	6.4 inches
Syrian	200,000 sq mi	Asia	< 5 inches
Great Basin	190,000 sq mi	North America	6–12 inches
Chihuahuan	175,000 sq mi	North America	< 9 inches

GLOSSARY

aquifer: Water in rocks deep underground.

canal: A waterway made by people.

climate change: Changes in Earth's weather patterns over time that are leading to droughts, warmer temperatures, and other extreme weather events.

device: A tool to be used.

dilemma: A problem.

irrigate: To supply water to crops.

natural resource: Something in nature that is important to humans, like water, trees, or animals.

nomad: A person who moves from place to place, usually with their family.

overgrazing: Having too many animals feeding on a grassland.

spring: Water that comes up from the ground.

terrace farming: Planting crops at different levels on the side of a mountain or valley.

INDEX

WEBSITES

Due to the changing nature of Internet links, PowerKids Press has developed an online list of websites related to the subject of this book. This site is updated regularly. Please use this link to access the list:
www.powerkidslinks.com/hdylt/desert